NATURE'S WRATH
THE SCIENCE BEHIND NATURAL DISASTERS

THE SCIENCE OF
TSUNAMIS

LEON GRAY

Gareth Stevens
Publishing

Please visit our website, www.garethstevens.com. For a free color catalog of all our high-quality books, call toll free 1-800-542-2595 or fax 1-877-542-2596.

Library of Congress Cataloging-in-Publication Data

Library of Congress Cataloging-in-Publication Data

Gray, Leon, 1974-
 The science of tsunamis / Leon Gray.
     p. cm. — (Nature's wrath: the science behind natural disasters)
 Includes index.
 ISBN 978-1-4339-8668-0 (pbk.)
 ISBN 978-1-4339-8669-7 (6-pack)
 ISBN 978-1-4339-8667-3 (library binding)
 1. Tsunamis. I. Title.
 GC221.2.G72 2013
 551.46'37—dc23

                    2012034535

First Edition

Published in 2013 by
Gareth Stevens Publishing
111 East 14th Street, Suite 349
New York, NY 10003

© 2013 Gareth Stevens Publishing

Produced by Calcium, www.calciumcreative.co.uk
Designed by Simon Borrough and Nick Leggett
Edited by Sarah Eason and Vicky Egan
Picture research by Susannah Jayes

Photo credits: Cover: Top: Shutterstock: Mana Photo; Bottom (l to r): Shutterstock: Steven Collins, Ben Heys, Paul Prescott, A.S. Zain. Inside: Dreamstime: Airnie 16–17tc, Pathathai Chungyam 41br, Jdanne 16–17bc, Linda Lim 37, David Snyder 31b, 36tr; FEMA: Dan Stoneking 44b; International Tsunami Information Center: 33c, 38c, 39c; NOAA: NGDC 28c; Shutterstock: 23bl, 3777190317 29b, David Aleksandrowicz 10cr, Darren K. Bradley 9b, Fouquin Christophe 4cr, Songquan Deng 18tr, Dvarg 25b, EpicStockMedia 21, 24tr, Gl0ck 7, iPhotos 6tr, JustASC 41tr, Dariush M. 13t, Dudarev Mikhail 20tr, Dale Mitchell 14c, Melody Mulligan 11b, Romchew 12c, Tae208 22c, Christian Vinces 1b, 19cl, Visdia 8cr, Wildnerpix 15c, Winconsinart 26tr, A.S. Zain 5t; US Navy: Photographer's Mate 3rd Class Tyler J. Clements 34tr, 35cl, Photographer's Mate 3rd Class Jacob J. Kirk 45bl, Mass Communication Specialist 3rd Class Alexander Tidd 30t, 43tl, Lance Cpl. Garry Welch 42–43c, Photographer's Mate 3rd Class M. Jeremie Yoder 32r; Wikipedia: David Rydevik 1c, 27.

Printed in the United States of America

CPSIA compliance information: Batch #CW13GS: For further information contact Gareth Stevens, New York, New York at 1-800-542-2595.

CONTENTS

What Is a Tsunami?   4

Chapter One: Underwater Forces
Violent Forces   6
Moving Plates   8
The Force of Earthquakes   10
Violent Volcanoes   12
Devastating Landslides   14
Real-Life Science: Lituya Bay, 1958   16
Real-Life Science: Super Tsunami   18

Chapter Two: Making Waves
What's in a Wave?   20
Tsunami Waves   22
Wave Motion   24

Chapter Three: Hitting the Shore
First Sightings   26
Total Destruction   28
Emergency Efforts   30
Real-Life Science: Indian Ocean, 2004   32
Real-Life Science: Thousands Dead   34

Chapter Four: Tsunami Warning
Tsunami Alert   36
Tsunami Science   38
Real-Life Science: Japan, 2011   40
Real-Life Science: Global Alert   42
The Challenge Continues   44

Glossary   46
For More Information   47
Index   48

WHAT IS A TSUNAMI?

A tsunami is the name given to a series of huge waves in the ocean. A tsunami occurs when a violent force on the seabed, such as an earthquake, shakes up the water in the ocean. This force creates a series of waves that move across the ocean at speeds of up to 600 miles (960 km) per hour—that is as fast as a passenger jet. The deadly waves crash onto land with an enormous force, causing widespread destruction.

The size of tsunami waves are almost unimaginable unless seen in real life. These massive walls of water can tower high over buildings. This artist's impression shows what a real-life tsunami might look like.

Ripple Effect

The effect of an violent undersea movement is similar to dropping a pebble in a pond, but on a much bigger scale. A pebble creates waves, which spread out as ripples across the surface of a pond. When a violent force shakes up the ocean to create a tsunami, the waves spread out in a similar way. In deep water, tsunami waves move very quickly, but are not very big. As they approach shallow water near the shore, the waves slow and grow into an enormous wall of water. The tsunami crashes into the coast with incredible force. The waves sweep away everything in their path, including villages, towns, and people.

Sumatra was one of the countries that took the full impact of the 2004 tsunami. Enormous waves up to 30 feet (9 m) high hit the country's shores, leaving a scene of devastation in their wake.

What's in a Name?

The word *tsunami* comes from the Japanese words *tsu*, which means "harbor" and *nami*, which means "wave." Since the effects of a tsunami are not felt far out on the deep ocean, Japanese fishermen would often not realize a tsunami had struck the shore until they returned to shore to find their entire village completely destroyed.

WORLD'S WORST

The deadliest tsunami occurred on December 26, 2004, when a series of powerful waves spread out across the Indian Ocean. The tsunami devastated coastal regions in many countries, including India, Indonesia, and Thailand. More than 285,000 people were killed.

CHAPTER ONE

UNDERWATER FORCES

The causes of catastrophic tsunamis are violent forces on the seabed, such as earthquakes, landslides, and volcanoes. When these events occur under the ocean, an immense force is created. The force is so powerful that it drives up the water above it. The displaced water then forms a huge and enormously powerful wave.

When the seabed shifts during a disturbance, such as an earthquake, the movement shakes up huge volumes of water above it.

VIOLENT FORCES

To understand these forces, we need to look at Earth's structure. It has three main layers similar to those of an egg:
• The core
• The mantle
• The crust

The core is the yolk of the egg. It consists of molten metals, such as iron and nickel, and the temperature is incredibly hot—up to 8,500°F (4,700°C). The mantle is the white of the egg. It is made up of molten and semisolid rocks, at a temperature of up to 3,500°F (2,000°C). The crust is the outermost layer, like an eggshell. It is a thin layer of solid rock at Earth's surface.

WORLD'S WORST

Meteorites are giant pieces of rock and metal from space. Most burn up as they fall through Earth's atmosphere. Scientists think that around 3.5 billion years ago a meteorite hit the surface of the ocean, creating a giant tsunami. The waves swept around the world and destroyed much of the life on our planet.

The crust "drifts" on the molten mantle. It is this movement that can produce earthquakes and volcanic eruptions. When these happen in the oceanic crust, they can shake up the seawater and cause tsunamis.

Meteorites have the potential to create enormous tsunamis. If a large meteorite smashed into our oceans today, it would create another devastating tsunami like the one that hit the earth 3.5 billion years ago.

7

MOVING PLATES

Earth's outer layer, the crust, is broken up into large pieces, like the cracked eggshell on an egg. These large pieces of the crust are called tectonic plates. They fit together like the pieces of an enormous puzzle. The edges of the tectonic plates make lines of weakness in the crust. The weaknesses are created because tectonic plates push together, and also pull apart. The force of the pushing and pulling can be so great that it sometimes creates movements in Earth's crust.

This diagram shows magma rising up at the points where two or more of Earth's tectonic plates meet. These are areas where pushing and pulling takes place and where earthquakes may occur.

Bumping Together

Although Earth's tectonic plates are constantly moving against each other, the movements are usually very small, often just a few inches every year. Most of the time, these movements go unnoticed because they are too small to affect people. However, when two plates cannot move easily against each other, pressure builds up along the lines of weakness between tectonic plates. When the pressure becomes too great, the plates move suddenly against one another. The built-up pressure between the two plates is then violently released at the surface of the earth as an earthquake or a volcanic eruption.

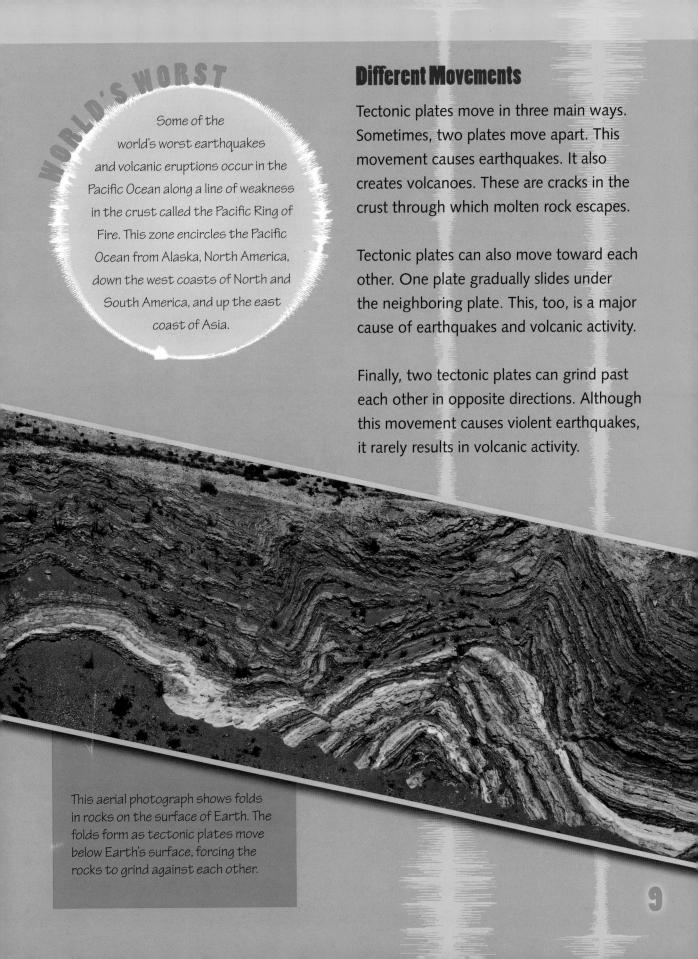

Some of the world's worst earthquakes and volcanic eruptions occur in the Pacific Ocean along a line of weakness in the crust called the Pacific Ring of Fire. This zone encircles the Pacific Ocean from Alaska, North America, down the west coasts of North and South America, and up the east coast of Asia.

Different Movements

Tectonic plates move in three main ways. Sometimes, two plates move apart. This movement causes earthquakes. It also creates volcanoes. These are cracks in the crust through which molten rock escapes.

Tectonic plates can also move toward each other. One plate gradually slides under the neighboring plate. This, too, is a major cause of earthquakes and volcanic activity.

Finally, two tectonic plates can grind past each other in opposite directions. Although this movement causes violent earthquakes, it rarely results in volcanic activity.

This aerial photograph shows folds in rocks on the surface of Earth. The folds form as tectonic plates move below Earth's surface, forcing the rocks to grind against each other.

THE FORCE OF EARTHQUAKES

Underwater earthquakes are the most common cause of tsunamis. At the bottom of the ocean, the tectonic plates move suddenly, creating an earthquake. The shock waves of energy made by the earthquake spread out into the surrounding water. The enormous release of pressure then creates the waves of a tsunami, and sets them on a collision course with the coast.

Just as huge rifts in Earth's surface are created by earthquakes on land, entire areas of the sea floor can be restructured during an undersea earthquake.

Under Pressure

Earthquakes occur when tectonic plates bump and grind against each other. The force of all this movement builds up in the surrounding rock. Eventually, this rock snaps apart under the enormous pressure. When rocks on the ocean floor slump downward, a dip forms on the seabed. Seawater rushes in to fill the hole, and the movement of water creates a tsunami. In coastal regions, an early warning sign of a tsunami is the sea pulling back from the coast to fill the dip. The sea floor is exposed. People know that the tsunami waves will follow in a few minutes and must move to higher ground quickly. When the rocks on the ocean floor push upward, this also shifts a huge volume of seawater. This creates the waves of a tsunami.

The Richter Scale

Scientists measure earthquakes using the Richter scale, which was developed by US scientist Charles F. Richter in 1935. It measures the amount of energy released during an earthquake. A hand grenade releases energy equivalent to 0.2 on the Richter scale. The earthquake responsible for the Indian Ocean tsunami of 2004 measured 9.2 on the Richter scale.

One of the deadliest earthquakes occurred in Lisbon, Portugal, on November 1, 1755. The shock waves ripped the city center apart. They also created an enormous tsunami in the Atlantic Ocean, which hit the city 30 minutes later. Survivors then faced a huge fire, which raged for days. In total, about 100,000 people lost their lives.

Eyewitnesses often speak of the pulling back of the ocean from the shore just before a tsunami hits, exposing large areas of sand.

VIOLENT VOLCANOES

A volcano is a mountainous buildup of land that contains hot material from within Earth's crust. There is a hole at the top of the volcano. When it erupts, hot gases and molten rock from deep inside spew out through the hole. The molten rock cools on the sides of the volcano. The eruption of a volcano in the crust at the bottom of the ocean can create an enormous force that causes a tsunami.

A cloud of hot gas billows from the top of this volcanic island. Tsunamis often occur around volcanic islands because the underlying rock is unstable.

Where Do Volcanoes Occur?

Volcanoes are common in regions where the tectonic plates are drifting apart or moving toward each other. Hot, molten rock, called magma, collects in magma chambers several miles beneath Earth's surface. As the plates move, enormous pressure builds up in the rock. Eventually, the rock cracks. The magma then rises up and bursts out onto the surface. Sometimes, gases also get trapped in the magma chamber. These heat up and explode as the eruption happens, making it even more violent.

Molten rock from exploding volcanoes can pour into the ocean. The force of falling rock piling into the water may generate devastating tsunami waves.

Volcanic Tsunamis

Sometimes, the upward force of the explosion from an underwater volcano moves enough water to create the waves of a tsunami. In other cases, after an underwater volcano has erupted, its magma chamber is left empty. The chamber collapses and seawater rushes in to fill the hole. This creates a surge of water in the form of tsunami waves traveling toward the shore. The tsunami waves created by underwater volcanoes can be as tall as the ones produced by the biggest earthquakes.

WORLD'S WORST

In 1883, the Indonesian island of Krakatoa was destroyed following a series of four enormous volcanic eruptions. The explosions were so violent that they were heard thousands of miles away in Perth, Australia. The collapse of the island created a series of massive tsunamis that struck land as far away as South Africa. A total of 36,000 people lost their lives in the aftermath of the eruption.

DEVASTATING LANDSLIDES

A landslide is the collapse of an area of land or ice such as a hillside or a glacier. Landslides of rock or ice along coastal areas can create tsunamis when enormous amounts of material crash into the ocean. Avalanches under the ocean can also create tsunamis. In both cases, the falling rocks displace a large volume of water, which creates the waves of the tsunami.

This massive chunk of ice fell from a glacier in Antarctica and crashed into the ocean. The force of falling ice such as this can displace enough water to create the waves of a tsunami.

The Fallout

Landslides often happen alongside earthquakes and volcanic eruptions, but the falling rock can often be more damaging than the earthquake or eruption itself. On November 18, 1929, an earthquake shook the Grand Banks off eastern Canada. The earthquake itself did not do much damage, but the landslide that occurred with the event created a tsunami that killed 28 people. People only knew a landslide had occurred because the falling rock damaged the underwater communication lines linking the United States and Europe.

Slipping and Sliding

Tsunamis that are caused by landslides are not always the result of earthquakes or volcanic eruptions. This makes them hard to predict. Any coastal area, or even an inland lake, could be hit by a tsunami caused by a landslide. For example, part of a nearby mountain could collapse and slide into the sea, or a mudslide could slip into a lake. In both cases, the debris from the landslide displaces the water, and waves spread out in in all directions. When the waves hit the coast, they can cause widespread devastation.

Many scientists think that a giant landslide will be the most likely cause of a future enormous tsunami, or "megatsunami." They suggest that the side of the volcano Mauna Loa, on Hawaii, might collapse. This landslide could trigger a megatsunami that would wipe out the city of Honolulu on the island of Oahu. Hundreds of thousands of people would lose their lives.

If part of Mauna Loa collapsed, hundreds of thousands of people on Oahu would lose their lives in the tsunami that would follow.

An enormous landslide, caused by an earthquake, created one of the tallest tsunamis in history. The giant tsunami struck Lituya Bay in the Gulf of Alaska, on July 9, 1958, shortly after 10:15 p.m. The landslide itself was caused by a huge earthquake in the Fairweather Range of mountains in Alaska. The force of the shock waves dislodged a massive amount of glacier ice in Lituya Bay, resulting in the huge tsumani.

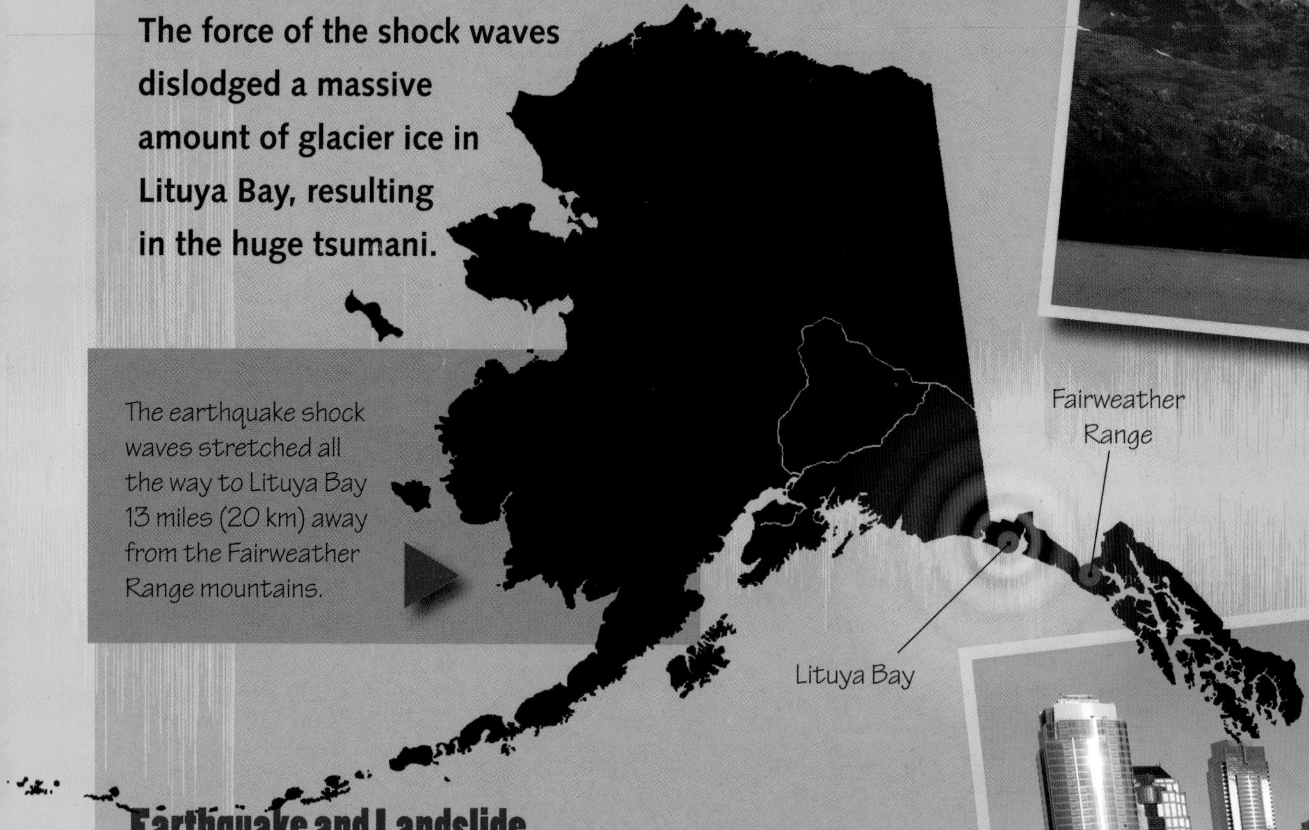

The earthquake shock waves stretched all the way to Lituya Bay 13 miles (20 km) away from the Fairweather Range mountains.

Fairweather Range

Lituya Bay

Earthquake and Landslide

A massive earthquake shook the Fairweather Range of mountains. It measured at least 7.9 on the Richter scale, and it sent out shock waves for thousands of miles around. The earthquake was so powerful that people as far away as Seattle, Washington, felt its effects. The earthquake triggered a massive landslide at Lituya Bay. A huge mass of rock and ice split off from one side of the bay. Many millions of tons of rock and ice plunged into the water.

This photograph shows the Fairweather Mountain range in Alaska—the center of the earthquake that resulted in the Lituya Bay landslide and subsequent tsunami.

The force of the earthquake in the Fairweather Mountains rocked skyscrapers in Seattle about 1,000 miles (1,600 km) south of Lituya Bay.

A Tsunami Follows

The impact of so much falling rock and ice generated a tsunami wave more than 1,720 feet (525 m) high. The wave swept out of the narrow bay at a speed of more than 130 miles (210 km) per hour. It traveled with such force that it stripped the land of vegetation, uprooting trees and huge rocks. It completely washed away the topsoil, leaving just the bare rock below. The wave then gushed into the Gulf of Alaska with incredible force, leaving the completely reshaped landscape of Lituya Bay behind it.

Husband and wife Bill and Vivian Swanson were asleep on their boat when the tsunami hit. Their boat, named Badger, was carried into the Gulf, riding the enormous wave like a surfboard:

"I saw the tops of trees on either side of the bay about 80 feet (25 m) below the boat as we were carried out into the sea. Badger broke up as the wave subsided, but we escaped in the dinghy and were picked up by a fishing boat about two hours later."

Survivors Speak

17

SUPER TSUNAMI

The tsunami at Lituya Bay was taller than the Empire State Building in New York City, which is 1,470 feet (450 m) tall from the base to the tip of its antenna. Compare the size of the building, shown center, to the boats in the foreground to see how huge the enormous wave was at Lituya Bay.

Life and Death

The tsunami claimed two lives: Orville Wagner and his wife, Mickey. The Wagners were moored in the bay on their boat and were hit by the full force of the wave. Amazingly, four other people who were also moored in the bay at the time survived the impact of the giant tsunami.

The Lituya Bay tsunami is the biggest wave in recorded history. Imagine a wave taller than the Empire State Building bearing down on you!

The Tsunami Disaster Unfolds

**JULY 9, 1958
10:15 p.m.**
A massive earthquake measuring 7.9 on the Richter scale strikes near the Fairweather Range Mountains.

10:17 p.m.
Shock waves from the quake vibrate in Lituya Bay, causing millions of tons of rock, ice, and debris to be dumped into the quiet waters.

10:18 p.m.
The landslide displaces a huge volume of water, setting in motion a tsunami that travels at more than 130 miles (209 km) per hour.

10:19 p.m.
The huge tsunami sweeps over the boat of Orville and Mickey Wagner, who are instantly killed.

10:22 p.m.
Howard Ulrich and his son Junior watch as the killer wave streams down the bay, building in height. The wave lifts the boat high out of the bay, snapping the anchor in the process. The boat then rides

Riding the Wave

As the wave hit, Howard Ulrich's boat was carried out on the crest of the wave toward the ocean. The backwash then carried it back into the bay. The water was calm enough for Howard and Junior to make their escape into the Gulf about 30 minutes later.

This artist's impression shows what the Swansons and Howard Ulrich and his son might have seen moments before the killer tsunami struck.

Howard Ulrich and his son, Junior, were in their boat, anchored in a small cove on the south shore of the bay. Both woke up at 10:15 p.m., when the shock waves from the earthquake violently rocked the boat. Minutes later, Howard heard a deafening explosion and saw a wall of water rushing toward his boat:

"The wave definitely started in Gilbert Inlet, just before the end of the quake. It was not a wave at first. It was like an explosion, or a glacier sluff. The wave came out of the lower part of the inlet, and looked like the smallest part of the whole thing."

the tsunami as the wave sweeps out of the bay toward the Gulf of Alaska.

10:23 p.m.
Husband and wife Bill and Vivian Swanson watch in fear as the giant wave lifts up their boat and sweeps

it out of the bay. Bill estimates they are more than 80 feet (25 m) above the tree line as the wave passes out of the bay and into the Gulf.

10:27 p.m.
Bill Swanson sees millions of gallons of

water and other debris pouring over the entrance to the bay.

10:30 p.m.
The Swansons' boat starts to break up. They abandon their ship and head away from the bay to safety.

10:35 p.m.
The backwash of the wave drags Ulrich's boat into the bay.

11:00 p.m.
The water in the bay gradually subsides, and Howard Ulrich steers the boat out of the bay to safety.

CHAPTER TWO

MAKING WAVES

The waves of a tsunami are completely different from normal waves you see on the surface of the ocean. They are a different shape, and they travel in a different way. Ordinary ocean waves are created by the pull of winds above the surface of the water. Tsunami waves are created by violent movements deep beneath the water surface.

The waves you can see on the surface of the ocean result from the constant push and pull of gravity and the wind.

WHAT'S IN A WAVE?

The wind creates normal waves on the surface of the ocean. As wind blows over the sea, the fast-moving air drags some of the surface water with it. The water rises because it cannot move as quickly as the air, but gravity pulls the water down again. It is the "tug-of-war" between gravity and the wind that causes the rise and fall of ocean waves.

Tidal Waves

Many people call tsunamis tidal waves, but these monster waves have nothing to do with the tides. Tides are created by the pull of gravity between Earth and the moon. As gravity from the moon pulls on the surface of the ocean, it creates currents of water that move up and down the shore.

WORLD'S ROGUES

Rogue waves are large freak waves. They form when strong winds and fast ocean currents join normal waves to create one giant wave. In 1995, a 90-foot (30 m) wave smashed into the ship *Queen Elizabeth II* during a hurricane in the Atlantic Ocean. The ship survived the onslaught, and no one was injured.

Even large ocean waves like this one can seem alarming. However tsunami waves are many, many times bigger and far more powerful.

TSUNAMI WAVES

Earthquakes, volcanoes, and landslides generate a huge force in the form of kinetic energy. The explosive force of an underwater earthquake or volcano must escape somewhere, so the kinetic energy transfers into the surrounding water. This causes a sudden, vertical displacement of an entire column of water, from the seabed right up to the surface of the ocean.

The crew of a fishing boat in the deep ocean may be completely unaware of a tsunami passing underneath.

Invisible Waves

The waves of a tsunami may be virtually undetectable in deep water. This is because they are only a few feet high, but spread over hundreds of miles. There are many stories of tsunami waves passing under ships on their way to shore without the crew even noticing them.

WORLD'S WORST

Krishnamurti Bedi was fishing the day the 2004 Indian Ocean tsunami hit the village of Pudhupettai in India. He felt a bump—but no more—as the tsunami wave passed under his boat. When Krishnamurti returned to shore, his village and his wife had disappeared, swept away by the massive tsunami wave.

The Wave Train

The huge volume of water displaced by an underwater movement rises above sea level. But gravity quickly pushes the water back down again. This sudden movement of a huge amount of water then creates the series of waves that form a tsunami. The tsunami waves spread out, like ripples on the surface of a pond. The rippling, successive waves that then descend relentlessly upon the shoreline are sometimes described as a "wave train."

Moving with Speed

In deep water, tsunami waves can travel extremely fast, at speeds of many hundreds of miles an hour. Tsunamis can move at such incredible speeds that they can cross the entire Pacific Ocean in less than a day. If a tsunami began near the coast of Japan at 9:00 a.m. in the morning, it could easily reach the eastern seaboard of the United States by the same time the following day! This travel speed is much faster than the usual large ocean waves created by the wind.

The wave train of a tsunami spreads through the ocean like the ripples on the surface of a pond.

Keep on Going

Although tsunami waves travel fast, they do not lose much of their energy as they move forward. Unlike normal waves, they also do not lose energy when they finally hit the coast. Instead, they plow into it with an incredible force and just keep on moving—across the land.

WAVE MOTION

Like all waves, tsunami waves have three main features: their height, their length, and their speed. These features change as the tsunami approaches the shore. As a tsunami wave nears the shore, it slows as the water drags along the ocean floor. The wave then builds in height to form a giant wall of water that bears down upon the land.

What Makes Up a Tsunami?

The amplitude of a tsunami wave is its height. The top of the wave is called the peak. The bottom is called the trough. In deep water, tsunamis have low amplitude, usually 2 or 3 feet (1 m) high. Normal waves have an amplitude of 6 feet (2 m).

The wavelength of the tsunami is the distance between one peak and the next. In the deep ocean, the wavelength of a tsunami is about 125 miles (200 km). The wavelength of a normal wave is around 330 feet (100 m).

The speed of a tsunami wave is much faster than a normal wave, too. In deep water, a typical tsunami wave may be traveling at 500 miles (800 km) per hour or more.

Large waves break on their journeys toward the shore. All waves have the properties of amplitude, speed, and wavelength.

WORLD'S WORST

In 1960, an earthquake struck off the coast of Chile in South America. Waves 80 feet (25 m) high hit the coast 15 minutes later. The tsunami reached Hawaii 15 hours later and then Japan—22 hours after the earthquake first took place!

Waves of Change

The features of the tsunami change as it enters shallow waters closer to shore. The wave slows to around 50 miles (80 km) per hour. The wavelength, the distance between the peaks, shortens to around 12 miles (20 km). The waves are now much closer together. But the amplitude, or height, of the waves increases. Each wave forms a wall of water that crashes into the coast with tremendous force.

The sheer force of the water from a tsunami has reduced this sea wall to rubble.

CHAPTER THREE

HITTING THE SHORE

As a tsunami hits the shallow waters close to the shore, the waves slow. Each successive wave builds into a huge wall of water that surges onto the land. The tsunami can hit the land in one of two ways, depending on whether the peak or the trough of the wave hits the coast first. An eyewitness looking out at the ocean will then see one of two things happen.

This artist's impression shows what a tsunami looks like when the peak of the wave hits the shoreline.

FIRST SIGHTINGS

If the peak hits first, an enormous wall of water appears on the horizon. The peak of the wave may foam, but it does not break, like a normal wave. Instead, the huge wall of water surges up onto the land. Water pours over the land with great force, destroying everything in its path. If the trough of a tsunami wave arrives at the shore first, the coastal water is sucked out to sea, exposing the seafloor. A few minutes later, the peak of the wave arrives and smashes into the coast. The waves of a large tsunami can rise to a height of 100 feet (30 m) or more and cause widespread destruction.

WORLD'S WORST

Even small tsunami waves can be very destructive. The 2011 tsunami that hit Japan produced 8-foot (20 m) wave surges 6,700 miles (10,800 km) away in California and Oregon. These caused a staggering $10 million worth of damage to buildings and highways.

More and More

Tsunamis consist of a series of waves that pound the shore again and again. The danger does not pass with the first wave. Often, it is the second or third wave that causes the most destruction.

This photograph was taken as a tsunami wave crashed into the coast of Krabi Province in Thailand during the 2004 Indian Ocean tsunami.

27

TOTAL DESTRUCTION

Tsunamis have a terrible power. They cause total destruction on coasts and islands. The locations affected by one tsunami may be thousands of miles apart, as the waves spread out from their source. It is not the force of the water alone that causes the damage—tsunamis pick up and carry forward anything in their path.

The surge of the water that followed the 2011 Japanese tsunami swept two ferries on to a road near Uranohama Port in Oshima.

Super Destroyer

The sheer power of a tsunami destroys everything in its path. The force of the water can uplift trees from their roots and strip the ground down to the bedrock. The power of the waves can wipe out coastal towns and villages, tearing houses from their foundations.

Tsunamis lift ships and boats high out of the water, smashing them into bridges and buildings. Fires break out at factories and oil refineries, and crops and livestock are destroyed. When the water subsides, the hardest hit areas can look like a war zone. The cost of the damage can run into many billions of dollars.

Loss of Life

Anyone caught in the path of a tsunami has little chance to escape as the huge volume of water floods the land. Tsunamis cause huge loss of life as people are trapped in fallen buildings and swept away in their cars. The sudden impact of a tsunami wave can take everyone by surprise, and often there is little warning of the impending danger. More than 285,000 people lost their lives following the Indian Ocean tsunami of 2004, in countries as far apart as India and Indonesia.

WORLD'S WORST

A tsunami that hit the eastern coast of Japan in 2011 caused a nuclear disaster when waves crashed into the Fukushima nuclear power station. More than 200,000 people had to be evacuated after radiation escaped into the surrounding area.

Many survivors of the 2004 Indian Ocean tsunami lost their homes as well as their relatives. A lot of families were forced to take shelter in tents in a makeshift refugee camp near Aceh, Indonesia, after the tsunami hit.

EMERGENCY EFFORTS

In the aftermath of a tsunami, people need a lot of help. In the short term, they need food and water, shelter, and medical aid. Longer-term efforts include rebuilding homes, and restoring services such as highways and hospitals.

Finding Survivors

The first stages in the emergency response are finding survivors and helping the injured to get medical attention as quickly as possible. People trapped in fallen buildings may have severe injuries, such as broken limbs. Others may have been swept miles out to sea as the waves of the tsunami retreated back to the ocean.

This aerial photograph shows the effect of the 2011 tsunami on the coastal village of Wakuya, Japan. The waves have flattened most of the buildings and swept boats far inland.

Rescue efforts are often hampered by the destruction left in the wake of a tsunami. Highways, bridges, and medical centers may have been destroyed. Communication may be difficult if telephone lines have been swept away. In the most remote or the hardest-hit areas, military troops often use helicopters to search for survivors from the air.

Finding Bodies

The task of recovering bodies is very important, not only for the loved ones, but also to prevent the spread of diseases, such as cholera. These diseases can quickly make large numbers of people sick. Rescue workers, troops, and volunteers scour the flattened land in search of victims buried under rubble or mud. The victims must be identified so that relatives can be told.

Rebuilding Lives

The next stage in the rescue effort is providing food and shelter for the survivors. Countries rely on aid agencies, charities, and foreign governments to help provide food, drinking water, clothes, and other essential supplies. Evacuation centers provide temporary shelter such as tents for people who have lost their homes. Longer-term efforts include rebuilding houses and services such as bridges, hospitals, highways, and schools.

WORLD'S WORST

Most of the people who died in the 2004 Indian Ocean tsunami were women and children. Many women were killed by the series of waves, swept away as they waited by the shore for their husbands to return from fishing trips. Most of the children were simply too small and weak to survive.

Emergency workers in India bury the dead following the tsunami of 2004. Quick burials are vital to prevent the spread of disease.

REAL-LIFE SCIENCE
INDIAN OCEAN, 2004

The Indian Ocean tsunami of 2004 was one of the most destructive and deadly natural disasters the world has ever seen. The shock waves from a huge underwater earthquake set off a series of tsunami waves that swept across the Indian Ocean. More than 285,000 people lost their lives in this tragedy.

The Earthquake Strikes

The earthquake struck off the northwestern coast of Sumatra, on the morning of Sunday, December 26, 2004. It was the third worst earthquake in recorded history. Measuring between 9.1 and 9.3 on the Richter scale, it released the same energy as 22,500 atomic bombs. The shock waves of the enormous earthquake were felt as far away as South Africa and Mexico.

This photograph, taken from a helicopter by US soldiers days after the 2004 tsunami, shows the complete destruction to the coastline of Aceh in Sumatra, Indonesia.

This map shows where the 2004 earthquake was most powerful, directly off the northen coast of Sumatra. The island was one of the countries most devasted by the tsunami.

epicenter

Sumatra

The Wave Train Hits

The earthquake set in motion a wave train of powerful tsunami waves in the Indian Ocean, which devastated coastal regions as far away as India and Somalia in Africa. Sri Lanka, Sumatra, and Thailand were some of the hardest-hit regions. They bore the full force of the waves that surged at great speed across the ocean and reached heights of 115 feet (35 m) as they hit land.

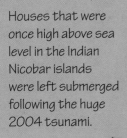

Houses that were once high above sea level in the Indian Nicobar islands were left submerged following the huge 2004 tsunami.

Journalist Barry Jones was asleep when the tsunami hit the town of Unawatuna in Sri Lanka:

"Suddenly, water was gushing under the door and through the windows of my room... The current was so strong it carried me along—past motorbikes, cars, trees, and fridges. Eventually I managed to grab hold of a pillar on a building and held on until the water subsided."

Survivors Speak

Hitting Sumatra

The first waves hit the coast of Sumatra about 25 minutes after the earthquake. Banda Aceh, a town on the northwestern coast, was flattened by the tsunami, which demolished buildings, ripped up vegetation, and flooded everything up to more than 2.5 miles (4 km) inland. About 90 minutes after the earthquake, the tsunami hit the coast of Sri Lanka. It swept away boats, buildings, and people. Thailand was hit at the same time, with 16 to 32-foot (5 to 10 m) waves, taking sunbathing tourists by surprise. The waves hit India around two hours after the quake. It took 6 hours for the waves to reach countries on the eastern coast of Africa, such as Somalia and Kenya.

THOUSANDS DEAD

It is estimated that at least 285,000 people in more than 14 countries lost their lives as a result of the 2004 earthquake and tsunami. In Indonesia alone, at least 170,000 people died in the disaster, while 35,000 were killed in Sri Lanka. Thousands more died in India and Thailand. In addition to the local people, many tourists were killed on the beaches of popular holiday destinations in the region.

No Warning

The lack of a tsunami warning system in the Indian Ocean was one of the main reasons for the high death toll. The tsunami hit most people by surprise, and there was little time to escape.

A staircase is left standing in the ruins of a home in the city of Banda Aceh in Indonesia. It took many years and billions of dollars of aid to help people rebuild their lives following the 2004 tsunami.

Disaster in the Indian Ocean

DEC 26, 2004
12:58 a.m.
A massive earthquake occurs just off the northwestern coast of Sumatra, Indonesia.

1:07 a.m.
The Pacific Tsunami Warning Center station in Australia picks up the shock waves of the quake.

1:23 a.m.
A tsunami hits the northwest coast of Sumatra, completely flattening the coastal town of Banda Aceh.

2:23 a.m.
Tsunami waves hit the island of Sri Lanka and the east coast of India, destroying many coastal villages.

2:33 a.m.
Tsunami waves hit the coast of Thailand, first in Phuket and later Khao Lak. Successive waves strike every 10 minutes.

3:00 a.m.
Waves batter the southern tip of India and the Maldives.

6:00 a.m.
Somalia and Kenya on the east coast of Africa are hit by the tsunami.

6:31 a.m.
The death toll rises to 500 in Sri Lanka. Thailand reports 55 deaths and hundreds more injuries.

The Aftermath

Countries around the world rallied to provide help in the aftermath of the Indian Ocean tsunami. Governments gave billions of dollars of aid, while the public gave billions more to aid organizations and charities to help the victims of the disaster. It took many years for people to recover. The relief effort helped families rebuild their homes and lead normal lives again. The money also helped build a new Indian Ocean tsunami warning system, which was launched in 2006.

The United States Agency for International Development (USAID) rushed to the aid of the victims of the tsunami.

Survivors Speak

Mary was one of thousands of tourists enjoying a beach holiday in Phuket, Thailand, when the tsunami struck on December 26, 2004. Looking out to the ocean, she noticed something strange:

"The sea was frothing, boats were bobbing up and down, and the water along the shoreline was being sucked back into the ocean. I'd seen a video in geography class about a tsunami that hit Hawaii in 1964, and I just knew a tsunami was on its way. Minutes later, an enormous wall of water appeared on the horizon. Everyone was screaming and running back to the hotel. I helped people inside, and we all headed up to the highest floor of the hotel. Then the tsunami surged onto the beach, and the wave swept away furniture, cars, and everything in its path."

7:42 a.m.
India reports 142 deaths and many still missing.

8:49 a.m.
Television reports start to confirm fears that the disaster is much worse. In India, 1,000 people are confirmed dead, 500 more in Sri Lanka, and 150 in Indonesia.

10:13 p.m.
The death toll rises around the world. Sri Lanka reports 1,500 fatalities. India deploys its navy to help with the relief effort, confirming at least 1,000 deaths.

12:00 p.m.
Officials estimate 26,000 people have died, with 12,000 in Sri Lanka alone, 5,000 dead in Indonesia, and 5,600 in India. Thailand reports 866 deaths, Malaysia 50, and the Maldives 43.

DEC 28, 2004
Death toll confirmed at 50,000 and rising. Thousands are still missing and feared dead.

JAN 1, 2005
Death toll confirmed at 150,000 and rising. Millions of people are left homeless. Survivors rely on the billions of dollars in aid pledged by countries around the world.

TSUNAMI WARNING

No one can prevent tsunamis from happening in the future, but there are things that can be done to prepare people for when they strike. Systems can be put in place to warn people when a tsunami is coming, so that they can take action and escape to a safe place. Around the world, a network of earthquake stations uses the latest technology to monitor the ground for earthquakes that could trigger tsunamis.

Survivors of the 2004 Indonesian tsunami know only too well the value of tsunami early warning systems.

TSUNAMI ALERT

In the world's oceans, a network of buoys has been installed to provide tsunami warnings. These are connected to sensors on the sea floor that measure changes in the water pressure that could indicate tsunami waves are forming. The buoys send signals to satellites in space, which then send the information to warning centers around the world.

Head for Safety

Hawaii is a group of islands that might be vulnerable to tsunami hits. Hawaii's officials have created a Tsunami Hazard Map that shows people the areas on the island that are safe to go to in the event of a tsunami disaster.

WORLD'S WORST

In 2012, an earthquake measuring 8.6 on the Richter scale struck off the coast of Sumatra. Unlike the 2004 event, the tsunami that followed did not cause a single death. Scientists picked up the shock waves and issued tsunami warnings. This gave local residents enough time to evacuate.

This tsunami-warning buoy is floating on the surface of the Andaman Sea near Phuket, Thailand. The system cannot stop tsunamis but can prevent the huge loss of life by issuing advance warnings.

TSUNAMI SCIENCE

The terrible devastation and loss of life in recent disasters, in the Indian Ocean in 2004 and Japan in 2011, have made people much more aware of the destructive power of tsunamis. They have also made scientists work hard to understand how tsunamis work and what can be done to save lives when they strike. Scientists now understand tsunamis better than ever.

Seismographs have been placed around the world. These machines pick up earthquake tremors. If any are picked up deep beneath the oceans, scientists know a tsunami may follow.

Learning from Disaster

Unfortunately, the best way to gather information about tsunamis is to study their actions and their effects when they strike. Scientists look at factors such as how far the waves can travel when they hit land, and how far above sea level the water reaches. Scientists compare satellite images of the land from before and after a tsunami, to measure these factors along coastlines.

WORLD'S WORST

The tsunami that hit the coast of Japan in March 2011 was one of the worst natural disasters in history. The warning systems worked fine—it was the sheer scale of the event that proved so deadly. The tsunami was so big it killed nearly 16,000 people.

Talking to Survivors

Scientists also learn from looking at evidence, such as eyewitness stories and historical accounts of past tsunami events. These records can tell them about the timing of the waves and their behavior once the waves reach the land. Evidence on the ground is also extremely useful. The destruction of vegetation and the scattered debris can tell scientists a lot about the behavior of the waves.

Scientists from the Pacific Tsunami Warning Center in Hawaii gather information from satellite images and buoys to issue warnings about tsunamis.

Tsunamis in Japan

The government in Japan has invested a lot of money to protect people from tsunamis. It has installed deep-sea tsunami detectors and warning systems so scientists can look out for earthquakes and predict tsunamis. Builders have constructed high seawalls to stop the waves and sturdy buildings to absorb the force of the water. Japanese children learn about tsunamis at school so they know what to do if they hear an alert. In every coastal village in Japan, there are evacuation routes that will lead people to safety.

REAL-LIFE SCIENCE
JAPAN, 2011

On Friday March 11, 2011, Japan was hit by a devastating tsunami. It was caused by the most powerful earthquake ever to hit the country. The earthquake struck about 42 miles (67 km) off the northeastern coast of Japan, deep beneath the floor of the Pacific Ocean. The initial shock measured 9.0 on the Richter scale—the most powerful earthquake Japan had ever witnessed—and two aftershocks caused even more damage. The tremors devastated bridges, buildings, and highways across the country.

epicenter

The earthquake that struck Japan in 2011 took place in an earthquake "hot spot" just off the country's coastline.

10-year-old Aito lived in Sendai, and was with his grandmother when the tsunami disaster struck:

"I grabbed my grandmother's hand and we ran up to the tsunami refuge behind our home. There, people screamed at us that the water was coming and we must run farther. With the water coming behind us, we ran up and up until we reached the cell phone tower. Below us, we saw that the whole of Sendai was flooded. It was unimaginable."

Survivors Speak

Feeling the Force

The earthquake uplifted a huge section of the seafloor by more than 26 feet (8 m). This triggered a massive tsunami. Within minutes, waves traveling at more than 500 miles (800 km) per hour hit the eastern coast of Japan's island of Honshu. The waves battered coastlines, breaking down seawall defenses and surging inland. The tsunami picked up any object in its path as it surged forward across the land.

Tsunami Surge

The tallest waves were 125 feet (38 m) high, but most were between 10 and 40 feet (3 and 12 m). Wave after wave of water surged over the seawall defenses, washing through harbors, ports, and coastal villages. The waves surged up to 6 miles (10 km) inland. About an hour after the earthquake struck, the seawater covered 180 square miles (470 square km). The worst hit areas were Kamaichi and the port town of Sendai, as well as several smaller islands.

The shocking disaster made headlines around the world. It was one of the worst natural disasters in decades.

Many children were orphaned during the Japanese tsunami, and countless parents lost their children in the disaster.

GLOBAL ALERT

Japan's tsunami also triggered a global warning, as the waves spread across the Pacific Ocean. People were evacuated all along the Pacific coast of North and South America, from Hawaii to Chile.

Relief Operations

Japan's government responded quickly, setting up evacuation shelters to house 300,000 people who lost their homes. It took many months to start rebuilding the villages, bridges, highways, and railways. In total, the tsunami caused $32.6 billion of damage. The after effects are still being felt, as Japanese scientists struggle to clear up the radioactive material that leaked when the tsunami waves battered the nuclear power station at Fukushima.

The powerful surge of water from the 2011 tsunami swept large ferries and fishing vessels into towns and villages along the coast of Japan.

The Earthquake and the Aftermath

MARCH 11, 2011
2:46 p.m.
An earthquake erupts under the sea off the northeast coast of Japan.

2:50 p.m.
Tsunami warnings are given out across the Pacific Ocean from Japan to the West Coast of the United States.

3:16 p.m.
The first waves hit the coast of Sendai.

3:55 p.m.
Waves more than 30 feet (9 m) high hit Sendai and sweep along the coast.

7:30 p.m.
Reports of casualties begin to be reported.

8:15 p.m.
A state of emergency is declared at the Fukushima nuclear power station.

9:35 p.m.
Four of Japan's nuclear power stations close down.

10:29 p.m.
Fukushima nuclear power station's cooling system fails.

10:48 p.m.
Residents within 1.5 miles (2.4 km) of Fukushima are told to evacuate the area.

MARCH 12, 2011
12:42 a.m.
The dam in Fukushima bursts, washing away five homes.

24-year-old teacher, Jessica Besecker, was in a classroom at Matsuiwa Junior High School in the village of Kesennuma on the day the tsunami struck:

"The school lies close to the coast but high above ground. So minutes later, when the tsunami rose, I could see the giant wave stretched across the horizon, its white crest advancing...Then I saw something catch fire. It was boats out in the harbor, and they were carrying the fire into the town on the wave."

Search and rescue teams looked for survivors in the rubble of buildings following the 2011 earthquake and huge tsunami.

2:06 a.m.
Radiation levels in Fukushima No. 1 nuclear power plant rise.

4:20 a.m.
Aftershocks and tsunamis continue.

12:00 p.m.
Refugees in northern Japan travel south. Millions of homes are without electricity.

4:00 p.m.
The death toll is now at least 900. Rescuers search for survivors amid the rubble.

4:19 p.m.
Fukushima is leaking radiation.

6:22 p.m.
An explosion at Fukushima causes the roof of a reactor to crash.

8:18 p.m.
People living within 12.5 miles (20 km) of Fukushima are told to leave the area.

MARCH 13, 2011
50,000 Japanese troops join the search and rescue operation. Three of Fukushima's four nuclear reactors now have major cooling problems.

MARCH 14, 2011
The death toll rises. Many rescue teams from other countries join the search for survivors.

MARCH 15, 2011
Aftershocks continue. Fire breaks out at the Fukushima nuclear power station.

THE CHALLENGE CONTINUES

Tsunamis are a constant threat for people living in coastal regions around the world. While many countries are well prepared for detecting earthquakes and issuing tsunami warnings, others are not. They need money from the global community in order to set up warning systems and plans that will educate the population about what they should do to stay safe. Without this, it will be difficult to avoid future disasters similar to the 2004 tsunami.

Changing Attitudes

Events such as the 2004 Indian Ocean tsunami and the 2011 Japan tsunami have changed the way people think about these disasters. No one can prevent the next tsunami, but more must be done to stop so many fatalities and help the survivors rebuild their lives.

Aid workers and survivors of a tsunami that hit American Samoa in 2009 worked together to clean up the piles of debris in the village of Pago Pago.

Poverty and Tsunamis

Some of the world's poorest countries cannot afford to spend money on advanced warning systems that use expensive satellite technology. Even if they know a tsunami is on its way, it is hard to warn people living in remote coastal villages. Most people living there still do not know what do when they are told a tsunami is approaching. It is up to their governments to educate them about the dangers of tsunamis. People should be told how to escape to high ground, or travel a good distance inland, to be safe when a tsunami hits.

A Helping Hand

Many people are shocked when they see the images of tsunamis on their televisions. They do not realize the scale of the destruction caused by these giant waves. Governments worldwide offer help to countries affected by tsunamis. Thankfully, there are also charities, such as the International Red Cross, that supply food, medicine, and shelter to people devastated by these deadly natural disasters.

People from a village on the Indonesian island of Sumatra desperately needed food, water, and other humanitarian aid following the 2004 tsunami.

GLOSSARY

aftermath: the time after a disaster, such as a tsunami, when tasks include finding survivors and cleaning up

aftershock: the earthquake tremors that follow the main event. They can be almost as severe as the first earthquake.

amplitude: the height of a wave, from its trough to its peak

cholera: a disease that causes severe diarrhea and which can kill

current: powerful movement in a body of water

debris: rocks and other materials dumped on land or torn down during a natural disaster

displace: to move out of position

erupt: explode or bursts out

evacuated: moved away from an area of danger

gravity: the force that pulls objects toward the center of Earth

horizon: the imaginary line where the sky and the sea appear to meet

kinetic energy: the energy that an object has as it moves

magma chamber: the area inside a volcano where magma collects

molten: melted by great heat

peak: the top or crest of a wave

radiation: a poisonous substance created by radioactive activity, such as those in a nuclear power station

remote: far away

Richter scale: a scale used to measure the power of earthquakes. Charles Richter introduced the scale in 1935.

rogue wave: an unusually large ocean wave, which is a combination of one or more normal-size waves

sensor: an object that can feel and record movements

tide: the regular rise and fall of the oceans, caused by the gravity of the moon pulling on seawater

topsoil: the upper level of soil on the ground

tremor: a shaking movement in the ground

trough: the lowest part of a wave

FOR MORE INFORMATION

Books

Hawkins, John. *Tsunami Disasters*. New York, NY: Rosen Central, 2012.

Jeffrey, Gary. *Tsunamis and Floods*. New York, NY: Rosen Publishing Group, 2007.

Spilsbury, Louise and Richard. *Sweeping Tsunamis*. Chicago, IL: Heinemann Library, 2003.

Wade, Mary Dodson. *Deadly Waves: Tsunamis*. Berkeley Heights, NJ: Enslow Publishers, 2013.

Websites

Discover more about tsunamis in Hawaii and the Pacific Basin, with science, eyewitness accounts, and exhibits.
www.tsunami.org

The National Oceanic and Atmospheric Administration (NOAA) tsunami site features the science of tsunamis, advanced warning systems, community preparedness information, and ways to reduce their devastating impact.
www.tsunami.noaa.gov

Find out all about tsunamis, their impact on people, and the invaluable work of the Tsunami Warning System.
www.ess.washington.edu/tsunami/index.html

INDEX

A
aftershocks 40, 43
aid agencies 31, 35, 45
amplitude 24, 25

B
buoys 36, 37

C
cholera 31
core 6
crust 6, 8, 9, 12

E
earthquakes 4, 6, 8, 9,
    10, 11, 13, 14, 15, 16,
    17, 18, 19, 22, 24, 32,
    33, 34, 35, 36, 38, 39,
    40, 41, 42, 43, 44

F
fires 11, 28, 43

G
gravity 20, 23

I
India 5, 22, 29, 33, 34, 35
Indian Ocean tsunami
    (2004) 5, 11, 22, 29, 31,
    32, 33, 34, 35, 38, 44,
    45
Indonesia 5, 13, 29, 32, 34,
    35

J
Japan 23, 24, 39
Japan tsunami (2011)
    26, 28, 29, 30, 38, 40,
    41, 42, 43, 44

K
Krakatoa 13

L
landslides 6, 14, 15,
    16, 17, 18, 22
Lisbon, Portugal 11
Lituya Bay, Alaska (1958)
    16, 17, 18, 19

M
magma 8, 12
mantle 6
Mauna Loa 15
megatsunamis 15
meteorites 6, 7
moon 20
mudslides 15

N
nuclear power stations
    29, 42, 43

P
Pacific Ring of Fire 9
Pacific Tsunami Warning
    System 39

R
rescue efforts 30, 31, 34,
    35, 43
Richter scale 11, 16, 18,
    32, 36
rogue waves 20

S
satellites 36, 38, 45
shock waves 10, 11, 16,
    18, 19, 32, 34, 36
Somalia 33, 34
Sri Lanka 33, 34, 35
Sumatra 5, 32, 33, 34, 36,
    45

T
tectonic plates 8, 9, 10, 12
Thailand 5, 27, 33, 34, 35,
    37
tidal waves 20

V
volcanoes 6, 8, 9, 12, 13,
    14, 15, 22, 45

W
warning systems 34, 35,
    36, 37, 38, 39, 42,
    44, 45
wavelength 24, 25